Tons of Trucks!

FIRE TRUCKS

By Martin Harasymiw

Gareth Stevens
PUBLISHING

Please visit our website, www.garethstevens.com. For a free color catalog of all our high-quality books, call toll free 1-800-542-2595 or fax 1-877-542-2596.

Library of Congress Cataloging-in-Publication Data
Names: Harasymiw, Martin, 1978- author.
Title: Fire trucks / Martin Harasymiw.
Description: New York : Gareth Stevens Publishing, [2022] | Series: Tons of trucks! | Includes index. | Audience: Grades K-1.
Identifiers: LCCN 2020024598 (print) | LCCN 2020024599 (ebook) | ISBN 9781538262795 (library binding) | ISBN 9781538262771 (paperback) | ISBN 9781538262788 (set) | ISBN 9781538262801 (ebook)
Subjects: LCSH: Fire engines–Juvenile literature.
Classification: LCC TH9372 .H357 2022 (print) | LCC TH9372 (ebook) | DDC 628.9/259–dc23
LC record available at https://lccn.loc.gov/2020024598
LC ebook record available at https://lccn.loc.gov/2020024599

First Edition

Published in 2022 by
Gareth Stevens Publishing
111 East 14th Street, Suite 349
New York, NY 10003

Editor: Therese Shea
Designer: Deanna Paternostro

Photo credits: Cover Welland Lau/Shutterstock.com; p. 5 Gary Blakeley/Shutterstock.com; p. 7 Philip Arno Photography/Shutterstock.com; p. 9 LeoPatrizi/iStock/Getty Images Plus/Getty Images; pp. 11, 24 (ladder) Jerald LaOrange/Shutterstock.com; pp. 13, 24 (chainsaw) Gorgev/Shutterstock.com; p. 15 AlexSava/E+/Getty Images; pp. 17, 24 (hose) MoniqueRodriguez/iStock/Getty Images Plus/Getty Images; p. 19 Mike Brake/Shutterstock.com; p. 21 Troy Aossey/Stone/Getty Images; p. 23 Keith Muratori/Shutterstock.com.

Printed in the United States of America

CPSIA compliance information: Batch #CSGS22: For further information contact Gareth Stevens, New York, New York at 1-800-542-2595.

Contents

My Favorite Truck. 4

Fast Fire Trucks. 6

Fire Engines. 14

Helpful Trucks! 22

Words to Know 24

Index. 24

This is my toy truck.
It's a fire truck!

FIRE

This is a real fire truck!
It carries firefighters
to fires.

SAN MARCOS FIRE DEPT.
SAN MARCOS
COMMAND
TRUCK
5
T 5
5

These trucks
have loud horns!

Fire trucks
have ladders.
The ladder on top
is tall!

LFD
FREELS
L
TOWER
2

Fire trucks carry tools.
Firefighters use chainsaws!

This is a fire engine.
It carries a water tank.

FIRE DEPT.

It has long hoses.

Fire engines
pump out water.

ENGINE

Some carry
eight firefighters!

These trucks help!

BRIDGEPORT FIRE DEPT.
BRIDGEPORT FIRE DEPT.
LADDER FIVE
BFD
E-12

Words to Know

chainsaw

hose

ladder

Index

hoses 16

ladders 10

tools 12

water tank 14